The Science of Cursive Writing

A Teacher's Guide to Cursive Writing

Joyce S. Rankin

Copyright © 2025 Joyce S. Rankin. All rights reserved.

For information please write to: info@barringerpublishing.com

Barringer Publishing, Naples, Florida
www.barringerpublishing.com

Formatting and layout by Linda Duider

ISBN: 978-1-954396-89-0

Library of Congress Cataloging-in-Publication Data
The Science of Cursive Writing: A Teacher's Guide to Cursive Writing

Printed in U.S.A.

THE SCIENCE OF Cursive Writing
A TEACHER'S GUIDE TO CURSIVE WRITING

Introduction:

The Science of Reading has emerged as an essential educational framework, emphasizing evidence-based approaches to teaching literacy [1]. This method, born from the efforts of a Congressional panel of reading specialists led by Louisa Moats in 2000 [2], underscores the importance of phonemic awareness, systematic phonics instruction, fluency, vocabulary, and comprehension. A fascinating complement to this literacy instruction is the practice of cursive writing, which engages unique brain circuits and augments cognitive functions [3].

Cursive Writing and Brain Activity: Research reveals that cursive writing activates brain regions differently from print writing and typing, enhancing brain connectivity, memory formation, and learning [4]. This increased brain activity, particularly in the theta rhythm range, is crucial for optimal learning conditions [5]. Cursive writing also stimulates electrical activity in the brain's parietal lobe and central regions, which is vital for encoding added information and improving cognitive functions [6].

Benefits of Combining Reading and Cursive Writing: Integrating cursive writing with reading instruction can strengthen the neural pathways involved in literacy, thus improving spelling skills, memory recall, and overall literacy. By engaging in fine motor skills and muscle memory, cursive writing reinforces spelling patterns and enhances cognitive engagement. The physical act of writing in cursive helps students internalize the spelling and recognition of words, complementing the foundational skills acquired through reading [7].

Effective Implementation: Starting with printing allows students to develop the necessary fine motor skills and letter recognition before progressing to cursive writing. A recommended sequence for teaching cursive includes beginning with simpler strokes and gradually introducing more complex letters. Teacher-led instruction is crucial, offering expert guidance, interactive learning, and personalized support [8]. This approach ensures consistency, motivation, and effective technique, making cursive writing valuable to the educational toolkit.

Conclusion: Incorporating cursive writing into the curriculum enhances literacy skills and supports the cognitive development emphasized by the Science of Reading [9]. By combining these evidence-based practices, educators can provide a comprehensive and engaging learning experience that benefits students at all stages of their educational journey.

References:

1. National Reading Panel. (2000). Teaching Children to Read: An Evidence-Based Assessment of the Scientific Research Literature on Reading and Its Implications for Reading Instruction. Retrieved from National Reading Panel Report
2. Moats, L. C. (2000). *Speech to Print: Language Essentials for Teachers.* Brookes Publishing.
3. James, K. H., & Engelhardt, L. (2012). The effects of handwriting experience on functional brain development in pre-literate children. *Trends in Neuroscience and Education, 1*(1), 32-42.
4. Berninger, V. W. (2012). Evidence-Based, Developmentally Appropriate Writing Skills K-5: Teaching the Orthographic Loop of Working Memory to Write Letters So Developing Writers Can Spell Words
5. Longcamp, M., Zerbato-Poudou, M. T., & Velay, J. L. (2005). The influence of writing practice on letter recognition in preschool children: A comparison between handwriting and typing. *Acta Psychologica, 119*(1), 67-79.
6. Christensen, C. A. (2004). Relationship between Orthographic-Motor Integration and Computer Use for the Production of Written Text. *Developmental Neuropsychology, 26*(1), 213-232.
7. Cahill, S. M. (2009). Where does handwriting fit? Strategies to Support Academic Achievement. *Intervention in School and Clinic, 44*(4), 223-228.
8. Graham, S., & Harris, K. R. (2005). Improving the Writing Performance of Young Struggling Writers: Theoretical and Programmatic Research from the Center on Accelerating Student Learning. *The Journal of Special Education, 39*(1), 19-33.
9. National Institute of Child Health and Human Development. *Teaching Children to Read: An Evidence-Based Assessment.*

How to use this manual:

1. Introduce each part by explaining the start and follow-through of each shape. Allow ample time for students to understand and imitate the correct pencil movement, beginning with proper posture.

2. Allow sufficient practice time for students to reproduce each shape. As lessons progress, additional practice may include embedding cursive writing into other coursework throughout the day.

3. As students reproduce each task perfectly, they will move on to the next lesson using large-lined paper and large pencils to number 2 pencils with three-ring binder paper.

4. By purchasing this manual, you may copy pages for practice or, following instructions, pass out blank lined paper.

Moving to the second part of instruction, using No. 2 pencils and "College Lined" or "Three-holed" paper, students will continue to increase their writing speed until it becomes seamless with their thoughts as they write.

A Teacher's Guide to Cursive Writing

TABLE OF CONTENTS

PART ONE: SCOPE, SEQUENCE, AND AUTOMATICITY 7

PART TWO: PROPER POSTURE. 11

PART THREE: BASIC SHAPE INSTRUCTION 13

PART FOUR: LOWER-CASE LETTERS. 21

PART FIVE: CAPITAL LETTERS. 37

Part One

The concepts of **scope, sequence, and automaticity** are crucial in linking reading skills to cursive writing, ensuring a structured and effective learning process.

SCOPE

Scope refers to the comprehensive range of skills and content covered in instruction. In the context of reading and cursive writing, it involves:

- **Phonemic Awareness and Phonics:** Teaching the sounds that letters make and how they combine to form words.

- **Handwriting Techniques:** Instructing on proper pencil grip, posture, and stroke formation for cursive letters.

- **Vocabulary and Comprehension:** Expanding students' word knowledge and understanding of text.

- **Spelling and Writing:** Integrating cursive writing with spelling practice to reinforce learning through muscle memory.

SEQUENCE

The **sequence** is about the order in which skills are taught to build upon each other effectively:

- **Foundation First:** Start with print handwriting (*printing*) to develop fine motor skills and letter recognition.

- **Simple to Complex:** Introduce cursive writing, beginning with simple strokes and progressing to more complex letters.

- **Phonics to Fluency:** Move from phonemic awareness and phonics to more advanced reading and writing tasks as students gain proficiency.

- **Reinforcement and Practice:** Ensure regular practice and reinforcement of reading and writing skills to solidify learning.

AUTOMATICITY

Automaticity is the ability to perform tasks with little conscious effort, essential for fluent reading and writing:

- **Fluent Handwriting:** Cursive writing helps develop automaticity in handwriting, allowing students to write smoothly and efficiently.

- **Reading Fluency:** Automaticity in reading enables students to read quickly and accurately, which is essential for comprehension.

- **Integrated Practice:** Repeated cursive writing and reading practice can help develop automaticity in both skills, making the processes more natural and effortless.

LINKING READING AND CURSIVE WRITING

Educators can create a cohesive learning experience by addressing the scope and sequence of reading skills and cursive writing and fostering automaticity. This integrated approach helps students internalize spelling patterns, improve reading fluency, and develop a solid foundation for lifelong literacy.

> **Teacher-led instruction is essential to provide guidance, feedback, and personalized support, ensuring students build these skills effectively and confidently.**

Printing and letter/sound recognition should be taught before cursive instruction.

1. **Letter Recognition:** Printing helps children become familiar with the shapes of letters and their corresponding sounds, which is fundamental for reading.

2. **Motor Skills Development:** Learning to print first helps develop the fine motor skills needed for writing. These skills can then be transferred to cursive writing.

3. **Gradual Progression:** Starting with printing allows children to master the basics before moving on to the more fluid and continuous strokes of cursive writing.

References:
1. **National Association for the Education of Young Children (NAEYC).** *Developmentally Appropriate Practice in Early Childhood Programs Serving Children from Birth through Age 8.* This publication emphasizes the importance of foundational skills in letter recognition and phonemic awareness before advancing to more complex writing tasks like cursive writing.
2. **Graham, S., & Harris, K. R.** (2005). *Improving the Writing Performance of Young Struggling Writers: Theoretical and Programmatic Research from the Center on Accelerating Student Learning.* ("Improving the writing performance of young struggling writers ...") The Journal of Special Education, 39(1), 19-33. This study discusses the developmental readiness of middle school students for advanced writing tasks, highlighting the motor and cognitive skills necessary for cursive writing.
3. **Occupational Therapy Guidelines**: Many guidelines suggest that children develop sufficient fine motor skills by middle school, making this an appropriate time for cursive writing instruction.

Developmental Readiness:
- A firm understanding of how to print letters, the sounds they make, and how they form words should be evident before instruction in cursive begins.
- Some students with reading proficiency may be able to begin cursive writing by upper elementary grades.
- Introducing cursive writing requires consideration of developmental readiness. Middle school students have the reading, cognitive, and motor skills to learn cursive effectively.
- These references support that a structured progression from foundational reading skills to cursive writing, timed according to developmental readiness, enhances educational outcomes and student proficiency in written communication.

Assessment:

Pre-Instruction Assessment:
1. **Baseline Handwriting Sample**:
   - Have students write a short paragraph in print, noting their handwriting legibility, spacing, and overall neatness. The instructor should choose a topic that relates to something familiar to all students. If printing is their only written skill, accept that in the baseline writing.
2. **Reading and Spelling Sample**:
   - Conduct a short test on phonemic awareness, letter recognition, and basic spelling skills to establish a baseline. This can be as simple as students writing sentences spoken by the teacher.
3. **Cursive Knowledge Questionnaire**:
   - Ask students what they know about cursive writing, if they have had any previous experience with it, and their feelings or attitudes towards it.

Sample Questions and Text:
- Pre-Instruction Writing Prompt: "Write a paragraph about your favorite hobby."
- Reading and Spelling Questions: Include a mix of phonemic awareness tasks (e.g., identifying beginning and ending sounds), letter recognition, and spelling of familiar words.

Part Two

PROPER POSTURE

POSTURE (Key Issue)

Start with both feet on the floor approximately shoulder width apart. Place your left hand at the top of the paper. If you are left-handed, your right hand would be at the top of the paper.

Note the position of the pencil. A more oversized pencil may help with the beginning lessons. It is preferable to begin using the larger pencil and then move to a smaller, Number 2 pencil when all posture suggestions are adopted and comfortable.

Left-handed:

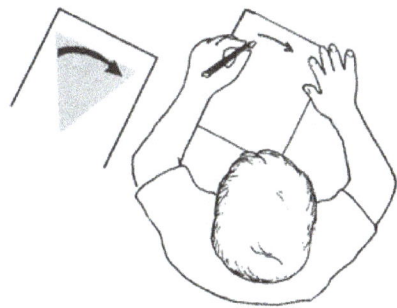

Right-handed:

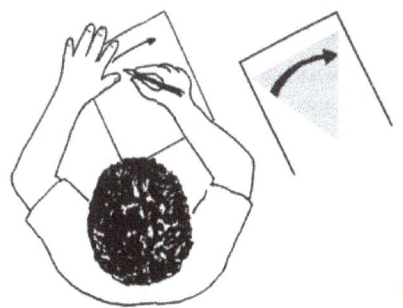

The correct sitting posture for handwriting.
1. Feet flat on the floor.
2. Thighs parallel to floor and knees at a 90-degree angle.
3. The back is straight, inclined towards the desk, and pivoted from the hips.
4. Forearms resting on the desk with elbows level with the desktop at 90 degrees.

The proper pencil grip for writing is known as the tripod grip.

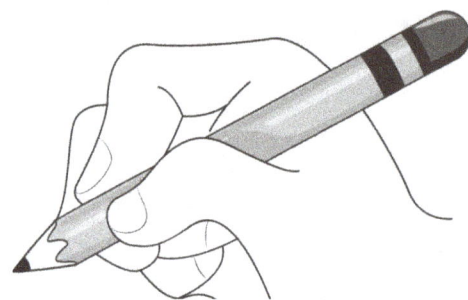

This grip helps maintain control and stability, which is essential for legible handwriting. Here is how to achieve it:

1. **Hold the Pencil with Thumb and Index Finger**: The pencil should be pinched between the thumb and index finger near the tip. The thumb should be opposite the index finger.
2. **Rest the Pencil on the Middle Finger**: The pencil should rest on the side of the middle finger, providing support and balance.
3. **Relax the Grip**: The grip should be firm enough to control the pencil but not too tight to cause strain. A relaxed grip allows for smoother writing movements.
4. **Proper Posture**: The wrist should be straight, not bent, and the hand should move smoothly across the paper. The other fingers should be relaxed and slightly curled.

TEACHER-LED INSTRUCTION:
- Provides expert guidance, interactive learning, personalized support, and consistent feedback.
- More effective than student-centered methods using booklets from the internet alone.

Part Three

BASIC SHAPE INSTRUCTION

THE CIRCLE

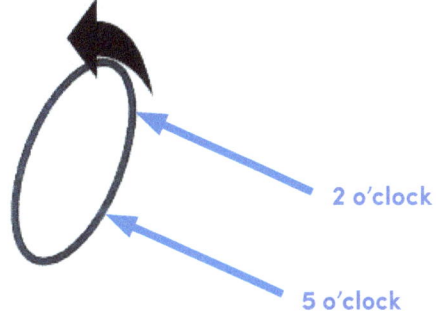

Start at 2 o'clock.
Move your pencil up and left to make an oval.
Finish making a "C" by stopping at 5 o'clock.

Students should have a large pencil and one page of three-lined paper.

Teaching instruction: The teacher leads the students in making a circle. The teacher then shows how the "O's" are connected across one line. Students can then work at their own speed while the teacher observes, making comments and/or corrections where necessary.

Check your posture. Stand and stretch if necessary. Roll your shoulders.

The teacher demonstrates how to make a continuous circle.

Instruction: Finish the top line and then **continue to complete the page**.

The top dot shows the student where to begin the circle. (Copy the next page or pass out lined paper.)

> Suggestion: When necessary, give students additional large-lined paper. *You may want to keep each student's papers in a folder for an evidence-based review.*

Continue with larger loops, allowing for additional practice time if necessary.

Instructions should continue for each shape. The teacher should observe and evaluate each student's work, and those who need remediation should be directed to review. Others can continue with additional shapes:

A Teacher's Guide to Cursive Writing

A Teacher's Guide to Cursive Writing 17

A Teacher's Guide to Cursive Writing 19

Review

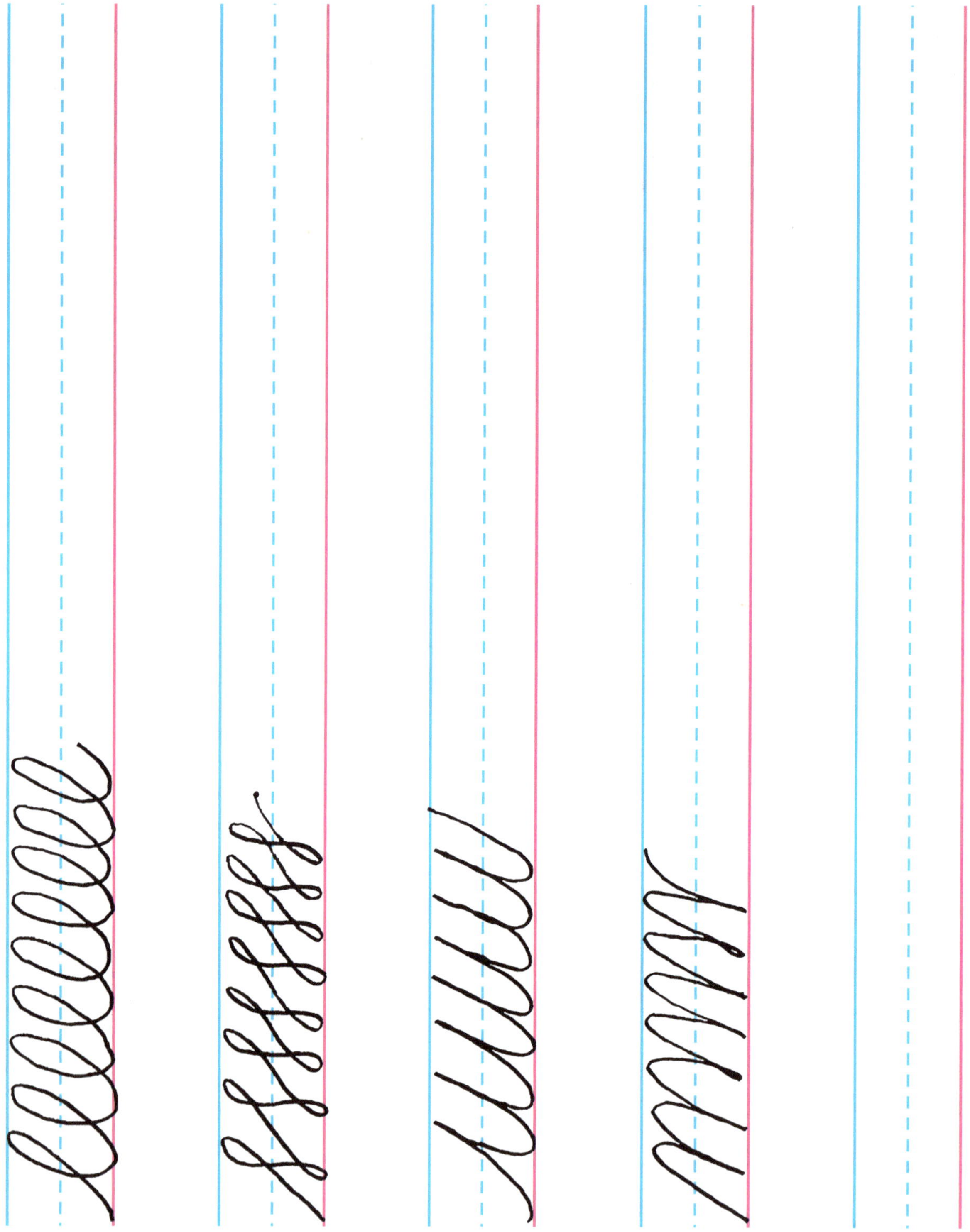

Continue with letter training as students are ready.

Teaching lowercase cursive letters in a structured sequence progressively helps students build their skills. Page 21 is a suggested sequence that starts with simpler strokes and gradually moves to more complex ones:

Part Four

LOWER-CASE LETTERS

Suggested Sequence for Teaching Lowercase Cursive Letters:

1. **Basic Strokes and Simple Letters:**
   - **l, i, t, u, w, e:** Begin with letters that have simple, straightforward strokes and loops. These letters help students get comfortable with basic cursive movements.

2. **Curved Strokes:**
   - **c, o, a, d, g, q:** Move to letters that involve curved strokes and more rounded shapes. These letters build on the basic strokes and introduce more fluid movements.

3. **Connecting Loops:**
   - **h, k, b, f, r, s:** Introduce letters that require connecting loops. These letters help students learn how to connect their strokes more seamlessly.

4. **Combination Strokes:**
   - **m, n, v, x, y, z:** Move on to letters that combine straight and curved strokes and more complex connections.

5. **Special Strokes:**
    - **j, p**: Finally, teach letters that require special strokes, such as descenders that dip below the baseline. **Below is a review of each letter and formation.**

Techniques for Teaching:

- **Demonstrate Each Letter:** Show students how to form each letter on the board, explaining the starting point, direction, and flow of each stroke.

- **Group Practice:** Practice writing each letter together, guiding students through the motions.

- **Individual Practice:** Provide worksheets with dotted or guided lines for students to practice writing each letter independently.

- **Reinforcement:** Review and reinforce proper letter formation regularly, ensuring students build muscle memory.

EXAMPLE LESSON PLAN:

1. **Warm-Up**: Start with warm-up exercises to loosen the hand and get students used to the cursive writing motion.
2. **Introduction**: Introduce a small group of letters that share similar strokes.
3. **Guided Practice**: Practice writing the letters together, focusing on proper technique and flow.
4. **Independent Practice**: Allow students to practice independently.
5. **Review and Feedback**: Check students' work and provide constructive feedback to correct errors.

This sequence and method should help students develop their cursive writing skills gradually and effectively. Suggested order:

- **Simple and Straight:** l, i, t, u, w, e
- **Simple Curves:** c, o, a, d, g, q
- **Basic Loops:** h, k, b, f, r, s
- **Challenging Letters:** j, m, n, p, y, z

Have students complete the line. If necessary, provide additional paper for practice.

Begin with a **large pencil** and **wide-lined paper**.

Simple and Straight

Simple and Straight   Simple Curves

lll　*ccc*　*ooo*　*aaa*　*ddd*

A Teacher's Guide to Cursive Writing

Simple Curves   Basic Loops

26 The Science of Cursive Writing

Basic Loops   Challenging Letters

A Teacher's Guide to Cursive Writing   27

Challenging Letters

Common Combinations

um um um

st st st

ra ra ra

on on on

Common Combinations

ay ay ay

at at at

ne ne ne

sh sh sh

ne ne ne

Common Combinations

um um um

se se se

ou ou ou

ie ie ie

re re re

Common Combinations

qu qu

ad ad

try try

A Teacher's Guide to Cursive Writing

The Science of Cursive Writing

CLASS EXERCISE

Class discussion: What words contain **Common Combinations**?
Print these as students add words to the discussion.

Have students write out the words in lowercase cursive.
Some examples:

under	only	have	write	quiet
street	hay	shine	set	bad
fast	pay	net	ease	baby
radio	hat	neat	you	hat

Optional review of lower-case letters:
- **Spelling Words**. The teacher says the words as students write them in cursive.
- Students can write words from other subject areas.
- Students write as the teacher speaks in simple sentences.
- Write the alphabet in continuous form.

Transitioning to "College" narrow lined paper and number 2 pencils.

Transition Timing:

1. **Mastery of Basic Skills**: Ensure that students have a strong grasp of both lowercase and uppercase cursive letters using larger pencils and paper. They should write with consistent letter formation, spacing, and legibility.

2. **Comfort and Confidence**: Look for signs that students are comfortable and confident in their cursive writing abilities. This includes being able to write entire words and simple sentences in cursive with ease.

3. **Introduction Phase**: Once students demonstrate proficiency with the basics, typically after several weeks to a few months of practice, you can begin the transition.

Transition Process:

1. **Gradual Introduction**: Start by introducing the narrower paper and #2 pencils during specific practice sessions rather than all at once. For example, have a dedicated "transition day" each week where students use the new tools.
2. **Guided Practice**: During these sessions, provide guided practice to help students adjust to the different feel and smaller writing space. Emphasize maintaining proper cursive form and neatness.
3. **Incremental Increase**: Gradually increase the frequency of using the narrower paper and pencils over a few weeks. Monitor students' progress and provide feedback to ensure they are adapting well.
4. **Full Integration**: Once students are comfortable using the new tools during dedicated sessions, fully integrate them into regular practice. Continue to provide support and corrections as needed.

Tips for Smooth Transition:

- **Encouragement and Support**: Encourage students and remind them that transitioning to standard tools is a sign of their progress and skill development.
- **Positive Reinforcement**: Use positive reinforcement to celebrate their successful adjustment to the new tools.
- **Adaptation Period**: Be patient and allow for an adaptation period. Some students may need more time to adjust than others.

By following this phased approach, you can help students smoothly transition from larger pencils and paper to narrower paper and #2 pencils, ensuring they continue to build on their cursive writing skills effectively.

Part Five

CAPITAL LETTERS

Techniques for Teaching:

- **Demonstrate**: Clearly demonstrate each capital letter on the board, explaining each stroke's starting point and direction.
- **Practice Together**: Have students practice each letter with you, repeating the strokes to build muscle memory.
- **Provide Worksheets**: Use worksheets with dotted or guided lines to help students practice forming each capital letter accurately.
- **Encourage Consistency**: Emphasize consistent practice and correct formation to ensure the students develop good habits.

Example Practice Session:

1. **Demonstration**: Show how to write the capital letter on the board.
2. **Guided Practice**: Students practice the letter along with you.
3. **Independent Practice**: Provide worksheets for students to practice independently.
4. **Review and Feedback**: Review students' work and provide feedback to correct any mistakes.

Following this structured approach, students can gradually build confidence and skill in writing capital letters in cursive.

Capital Letters in Cursive
The sequence for teaching cursive writing for capital letters is consistent with how you might approach lowercase letters. The key is to start with simpler strokes and gradually move to more complex ones.

Suggested Sequence for Teaching Cursive Capital Letters:

1. **Basic Curves and Simple Strokes**:
    - **C, O, Q**: These letters involve simple, rounded shapes that beginners can master.
2. **Straight Lines and Simple Loops**:
    - **I, J**: These letters involve straightforward lines with simple loops.
    - **U, V, W**: These letters are simple and involve straight lines with slight curves.
3. **Looping Capitals**:
    - **A, D, E**: These letters include loops and more complex strokes.
    - **H, K, L**: These involve tall strokes and loops.
4. **Mixed Curves and Angles**:
    - **B, P, R**: These letters combine straight lines and curves.
    - **M, N**: These letters involve multiple strokes and angular shapes.
5. **Complex Shapes**:
    - **G, S, Z**: These letters have more intricate shapes and often require more practice.
    - **T, F, Y**: These letters involve unique strokes and complex formations.

Basic Curves and Simple Strokes

Straight Lines and Simple Loops

A Teacher's Guide to Cursive Writing

Straight Lines and Simple Loops　　　　**Looping Capitals**

The Science of Cursive Writing

Looping Capitals

Mixed Curves and Angles

A Teacher's Guide to Cursive Writing

Mixed Curves and Angles

Complex Shapes

The Science of Cursive Writing

Complex Shapes

A Teacher's Guide to Cursive Writing

Use this page to review all letters. If students can reproduce this page confidently, move on to number 2 pencils and "College" (three-holed) paper. Students should now be able to produce the alphabet in a smaller format.

Practice, as needed, using "College" or three-ring binder paper. Ask students to produce daily work using cursive writing for additional and continuous practice.

Post-Instruction Assessment:

Cursive Handwriting Sample:
1. **Reading and Spelling Test**:
   - Re-administer the pre-course assessment incorporating phonemic awareness, letter recognition, and spelling tests to see improvements in these areas.
2. **Cursive Fluency Test**:
   - Assess students' ability to write cursive letters and words quickly and accurately, noting improvements in speed and accuracy.
3. **Self-Assessment and Feedback**:
   - Ask students to reflect on their progress by answering questions about their confidence in cursive writing, challenges, and what they feel they have improved on. Allow students to review folders of their work if applicable.
4. Include the post-instruction assessment in their folder.

Summary

Cursive Writing and Brain Activity:
- **Enhanced Connectivity:** Cursive writing activates more brain regions than print writing or typing, enhancing brain connectivity, memory formation, and learning.
- **Theta Rhythm Synchronization:** Writing in cursive synchronizes brain waves in the theta range, optimal for learning and memory.

Benefits for Spelling and Literacy:
- **Muscle Memory:** Cursive writing helps develop muscle memory, reinforcing spelling patterns and improving spelling accuracy.
- **Cognitive Engagement:** The physical act of cursive writing requires more cognitive engagement, aiding in spelling and word recognition.
- **Holistic Approach:** Combining reading instruction with cursive writing strengthens neural pathways involved in literacy.

Effective Instruction:
- **Scope and Sequence:**
    - Start with printing to develop fine motor skills and letter recognition.
    - Introduce cursive writing with more straightforward strokes first, progressing to more complex letters.
    - Integrate phonics, vocabulary, and spelling practice with cursive writing.
- **Teacher-Led Instruction:**
    - Provides expert guidance, interactive learning, personalized support, and consistent feedback.
    - More effective than student-centered methods using booklets from the internet alone.
- Finally, keep using cursive writing throughout all coursework. This practice will enhance speed and fluency.

References
1. National Handwriting Association. (n.d.). *Cursive Writing Instruction.* Retrieved from National Handwriting Association
2. Occupational Therapy Guidelines on Handwriting. (n.d.). *Best Practices for Handwriting Instruction.*
3. Various Educational and Teaching Resources. (n.d.). *Teaching Cursive Writing Effectively.*
4. Microsoft Copilot (2024). AI-Generated Guidance on Cursive Writing Instruction. Personal Communication.